The Knots Puzzle Book

Heather McLeay

Contents ...

Making a Start	2
Endless Knots	3
The Unknot	5
Crossing Numbers	6
Mirror Images	12
Prime and Composite Knots	14
The 3-Colour Test	18
Analysing Moves	22
Higher Crossing Numbers	25
Practical & Artistic Knots	27
More Advanced Ideas	35
Knot Classification	41
Solutions & Comments	47

KEY CURRICULUM PRESS
Innovators in Mathematics Education

Making a Start

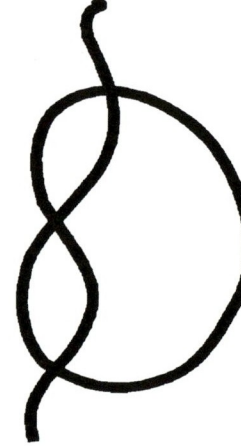

Let us start with a simple experiment. Take a piece of string in your hand and, without thinking, tie a knot in it. When you look at what you have done, you will probably find that you have tied a simple overhand knot like this one. Most people do.

This drawing shows the shape of the knot, but it is not a satisfactory diagram because it fails to show which strand passes over and which passes underneath. A good knot diagram is one which clearly shows exactly how the knot should be tied.

The drawing on the left shows a realistic rope, but the other two are easier to draw. All three styles show clearly which strand is on top.

Puzzle 1

The original shadow diagram could represent eight (2^3) different knots, because at each crossover there are two possibilities. Either strand could be on top. Four of the possible interpretations are shown below. Which are knotted and which are simply loops of rope?

A B C D

The Endless Knot

All these knots have three crossovers and were obtained by manipulating the original overhand knot. If the ends of a rope are loose, then it is always possible for a knot to become untied whilst it is being manipulated. For this reason a 'mathematical' knot is always studied with its ends joined together. Knots in this form, we shall call 'endless knots'.

Practically, an endless knot can be made by pushing the two ends into a piece of rubber tubing.

Since it does not matter where the join is, the convention is to show it as a continuous length of rope.

Once the ends of the rope are joined, it can no longer become untied by accident but more importantly and fundamentally, it remains unchanged in a real sense no matter how it is arranged or rearranged.

Joining the Ends

The obvious way to make an endless knot is to join the ends 'outside the loop'. However, what happens if we join the ends 'through the loop'?

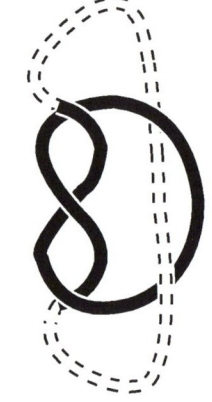

Ordinary overhand knot

Joined 'outside the loop'

Joined 'through the loop'

It is clear that these two methods of joining the ends do produce knots which are different. One has five crossovers and the other has three. As an experiment, use a real length of rope and join it so as to produce the knot joined 'through the loop'. Then try to rearrange it to have fewer crossovers. You will find that it cannot be done and therefore we can be certain that these two versions are indeed different.

Ordinary knots are turned into endless knots by joining 'outside the loop'. This agrees with the intuitive idea of pulling at the ends of any tangled loop of rope until it either comes undone or forms a 'lumpy bit' in the middle. A knot is then converted into its endless form by joining the two distant ends.

Puzzle 2

Careful inspection will show that there are two different ways in which a knot can be joined 'through the loop'. One produces a knot which always has five crossovers and one a knot which may have no real crossovers at all. Which is which?

A

B

The Unknot

Knots which can be rearranged to have no crossovers at all are called 'unknots'.

Puzzle 3

Rearrange these knots and decide which are unknots.

A B 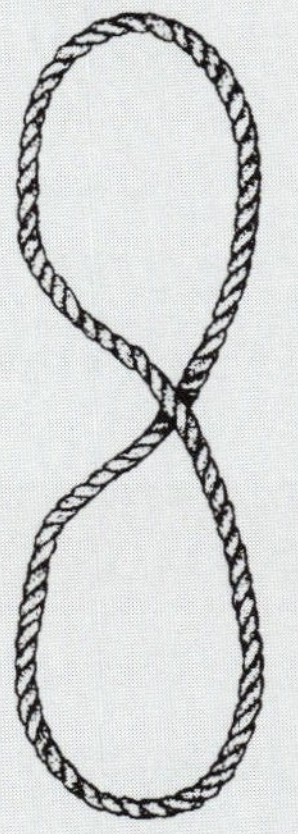 C 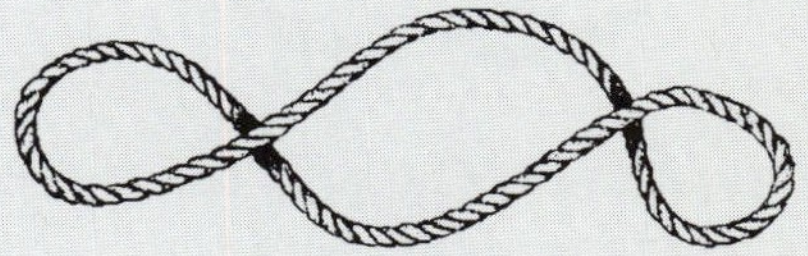 D  E  F  G  H

Crossing Numbers

In Puzzle 3 you found that many of the knots could be simplified to have fewer crossovers. However, each knot has a minimum number of crossovers and no further rearrangement will give less. This minimum number is known as the 'crossing number'.

An unknot has a crossing number of 0, because it can be arranged to have no crossovers at all.

Each of these knot diagrams has five crossovers. Try to determine the crossing number of each by mentally rearranging it or by tying and manipulating a real piece of rope. You should find that the crossing numbers are 0, 3 and 4 respectively.

Puzzle 4

Each of these knots has six crossovers. What are their crossing numbers?

An important part of the study of knots is to find techniques which allow us to be certain whether two knots are the same or different. We need to find those properties and characteristics of a knot which remain unchanged no matter how it is arranged or rearranged. Such a characteristic is known as an 'invariant' under those manipulations. The minimum crossing number of a knot is exactly the kind of invariant property that we are looking for.

Puzzle 5

What is the crossing number of each of these knots?

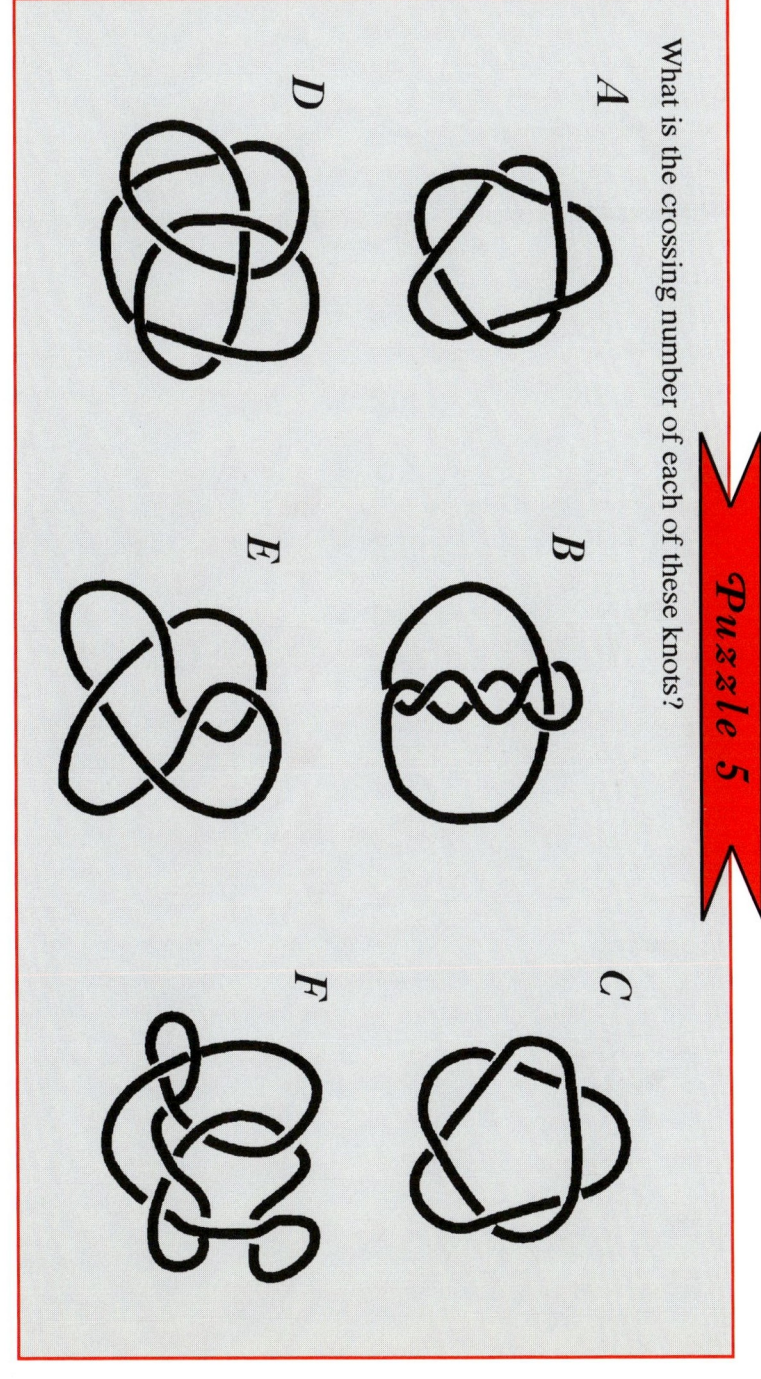

You will have noticed that all the knots in Puzzle 5 have different crossing numbers. We can therefore be certain that they are all different. However, as we shall see on page 10, the converse is not true. If two knots have the same crossing number, they are not necessarily the same knot.

Crossing Numbers 0 to 4

These knots appear to have crossing numbers of either 1 or 2 but, if you rearrange all of them mentally or manipulate them using a piece of real rope, you will soon see that they are actually unknots.

It is a fact that there are no knots which have crossing numbers of 1 or 2. However, there is one knot with a crossing number of 3 and one with a crossing number of 4. On page 12 we shall introduce the idea of whether a knot is the same as or different from its mirror image, but for the moment;

The only knot with a crossing number of 4 is called the 'figure-eight'.

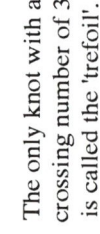

The only knot with a crossing number of 3 is called the 'trefoil'.

Puzzle 6

All four of these knots have 4 crossovers. When they are made into endless knots or simply have their ends pulled, which will be figure-eights?

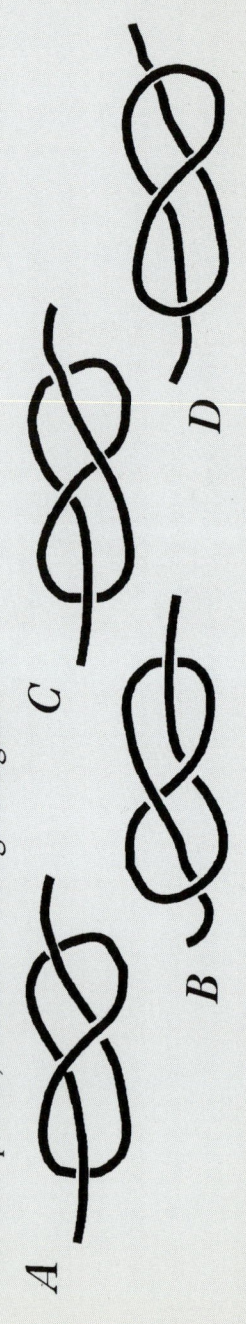

A B C D

Puzzle 7

By finding the crossing numbers of these knots, decide whether each knot is a trefoil, a figure-eight or an unknot.

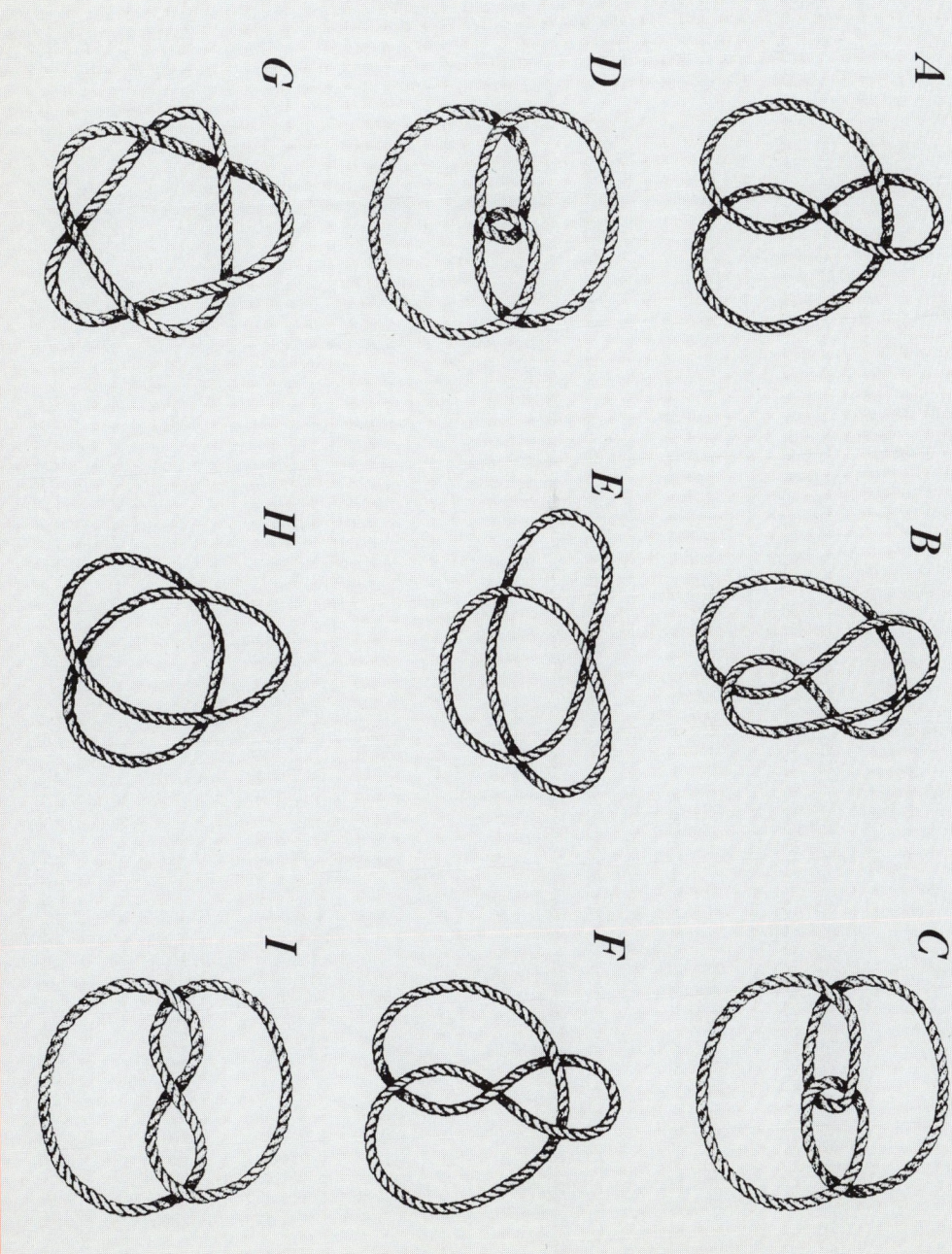

Crossing Number 5

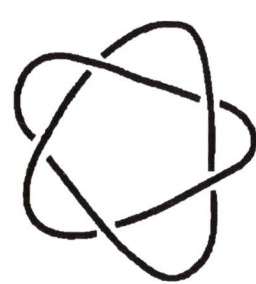

Each of these knots has a crossing number of 5 and they certainly appear to be different. The one on the left resembles a pentagon and it is called a 'pentoil'. The other is similar to the figure-eight knot but has an extra twist. These two knots seem to be different but how can we *prove* that they are?

There is a technique which we can use to distinguish between them and it requires that each crossover is labelled in some way. In these examples we have used letters of the alphabet.

Start at the arrow and follow the rope all the way round until you get back to it. This means that you will pass through each crossover twice, once along the strand on top and once along the strand underneath. Record the sequence of crossovers which you pass.

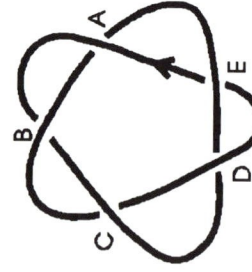

Example 1:
The sequence of crossovers is
A B C D E, A B C D E

Example 2:
The sequence of crossovers is
A B E D C, A B C D E

A characteristic of the pentoil knot is that the letters are repeated twice in the same order. If this does not happen, the knot is not a pentoil. We can therefore use this test to distinguish between them even when they are arranged in less familiar forms.

Look at these knots and try to work out in your mind which is a pentoil. Then check your answer by finding the sequence of crossovers.

Example 3:   The sequence of crossovers is
A B C D E, C B A D E.
The order does not repeat.

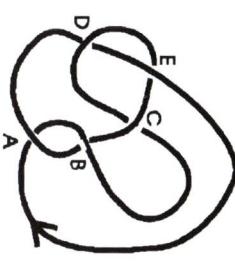

Example 4:   The sequence of crossovers is
E A B C D, E A B C D.
The order does repeat.

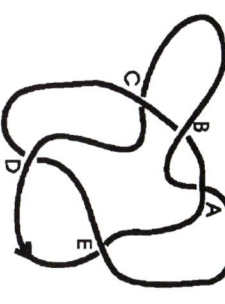

Hence Example 4 is a pentoil and Example 3 is the figure-eight form. Another way to confirm these results is to use a real piece of rope and try to rearrange knots 3 and 4 into the standard forms.

Puzzle 8

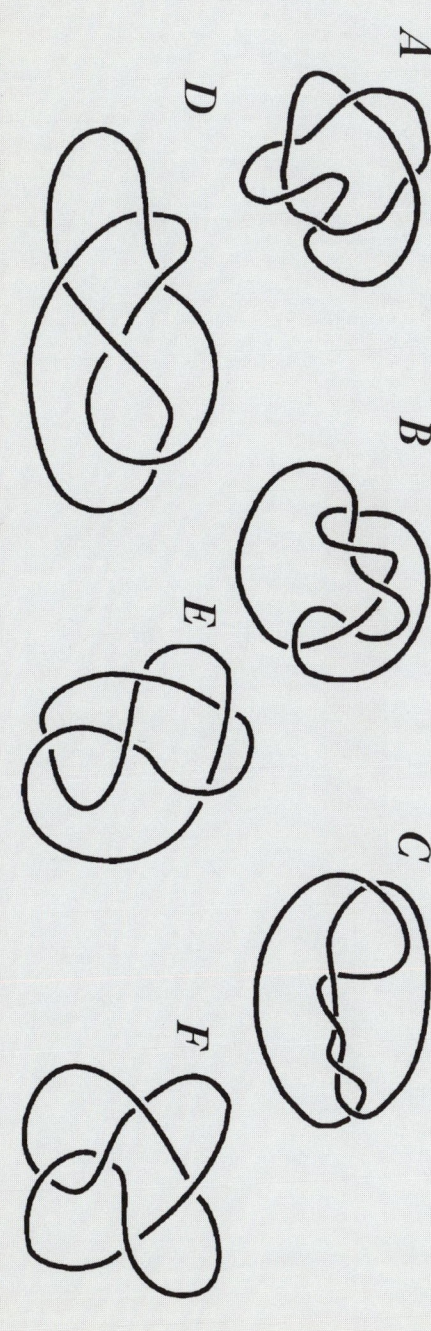

Label and find the sequence of crossovers for each of these knots. Then decide which are pentoils.

11

Mirror Images

Suppose that we look at a knot in a mirror. Is it changed or does it stay the same?

TREFOIL TREFOIL (mirrored)

The mirror image of a trefoil certainly seems to be a trefoil and it has already been stated that there are no other knots with a crossing number of 3. The common sense reaction is to say that the mirror image of a knot is simply the back view, the view from the other side. However this is absolutely not true! For a trefoil knot, the back and front views are identical whereas the knot and its mirror image are different. An easy way to appreciate this is by tying both kinds of trefoil using real rope and then turning them over. You cannot convert one into the other without retying it.

Anticlockwise Clockwise

While it is certain that they are different, it is very difficult indeed to describe the difference. The ideas of left-handed and right-handed can be used, or clockwise and anti-clockwise but we still have to define which name should be associated with which version of the knot. The directions of the arrows in these diagrams have been chosen quite arbitrarily.

Puzzle 9

Which of these knots match the clockwise diagram above?

A B C

Let us now examine the figure-eight knot and its mirror image.

FIGURE-EIGHT / FIGURE-EIGHT

Once again, the mirror image of this knot is not the same as the view from the other side. Convince yourself of this fact by tying a figure-eight knot in some real rope. Turning it over produces the same knot, not its mirror image. However, and this is part of the fascination of the study of knots, a figure-eight knot can be manipulated and rearranged into its mirror image form. The two versions are not distinct. Simply follow the instructions below.

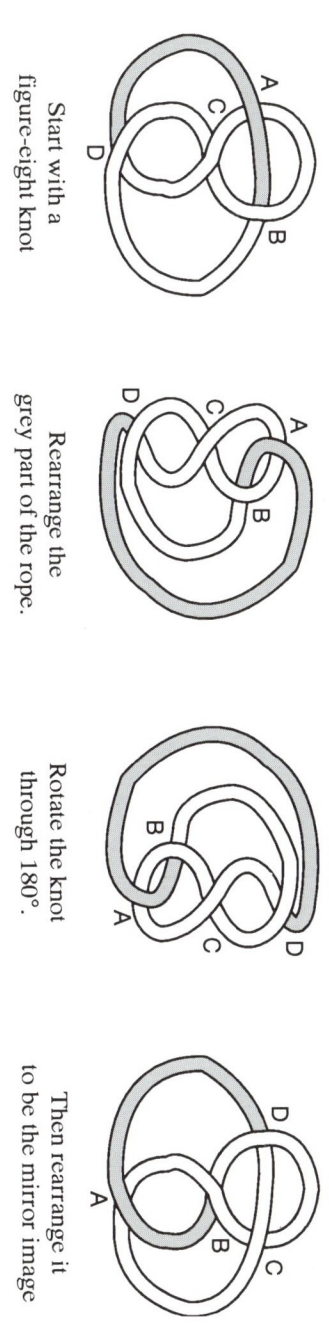

Start with a figure-eight knot

Rearrange the grey part of the rope.

Rotate the knot through 180°.

Then rearrange it to be the mirror image

It is a curious fact that turning over an overhead projector slide diagram of a knot produces its mirror image whereas turning over the knot itself does not. One of the trickiest problems in knot theory is to prove whether a knot and its mirror image are distinct or not. With more complicated knots it can be difficult to decide just by trying to rearrange them.

A better and certain method of proof is to make use of the Kauffman polynomial, another invariant of knots. These polynomials are beyond the scope of this book, but they can resolve the problem. With their aid it can be shown that the mirror images of knots with odd crossing numbers are distinct whereas some knots with even crossing numbers have distinct mirror images and others do not.

13

Composite Knots

A composite knot is a knot that is made up of two or more simpler knots.

Knot A Knot B Knot C Knot D

Each of the four knots on this page has a crossing number of 6 but they can clearly be classified into two different types.

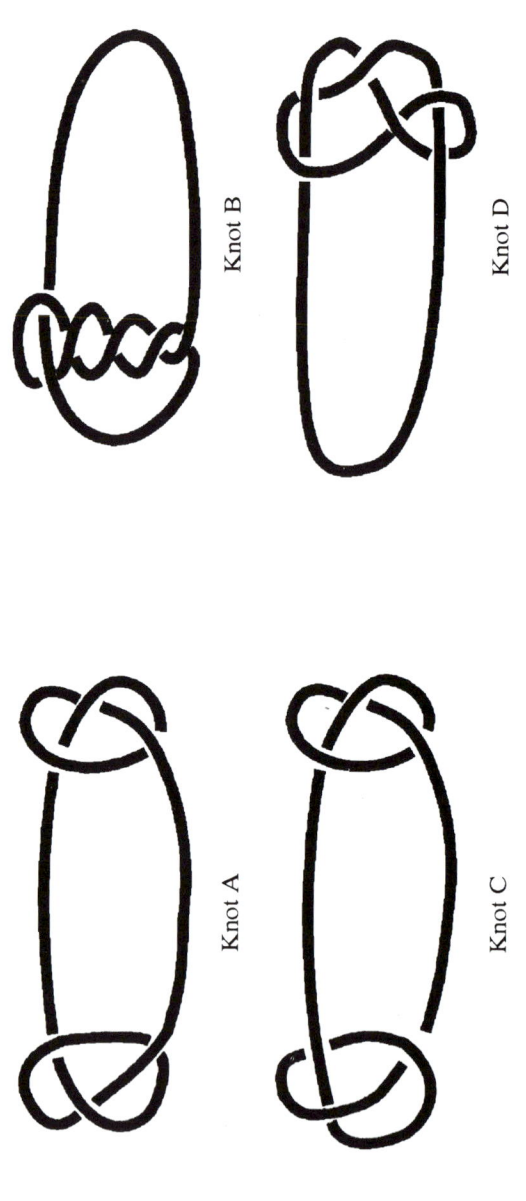

Knot A Knot B Knot C Knot D

Knots A and C can be teased out into a form where they are each made up of two simpler knots linked together. Knots which have this property are known as 'composite knots'. Knots B and D differ in that they will not separate out into anything simpler. They are therefore examples of 'prime knots'.

It is recommended that you convince yourself that neither of the prime knots can be separated out by tying them in a real piece of rope.

Reef Knot or Granny

The two knots which most people have heard of, whether they are interested in knot theory or are simply interested in tying a parcel securely, are the reef knot and the granny. It always seems surprising that such a small variation in the method of tying the knot produces either one which pulls tight and holds firm or one which slips easily. Most of us have to learn the hard way which is which.

We are now in a position to understand the difference between them.

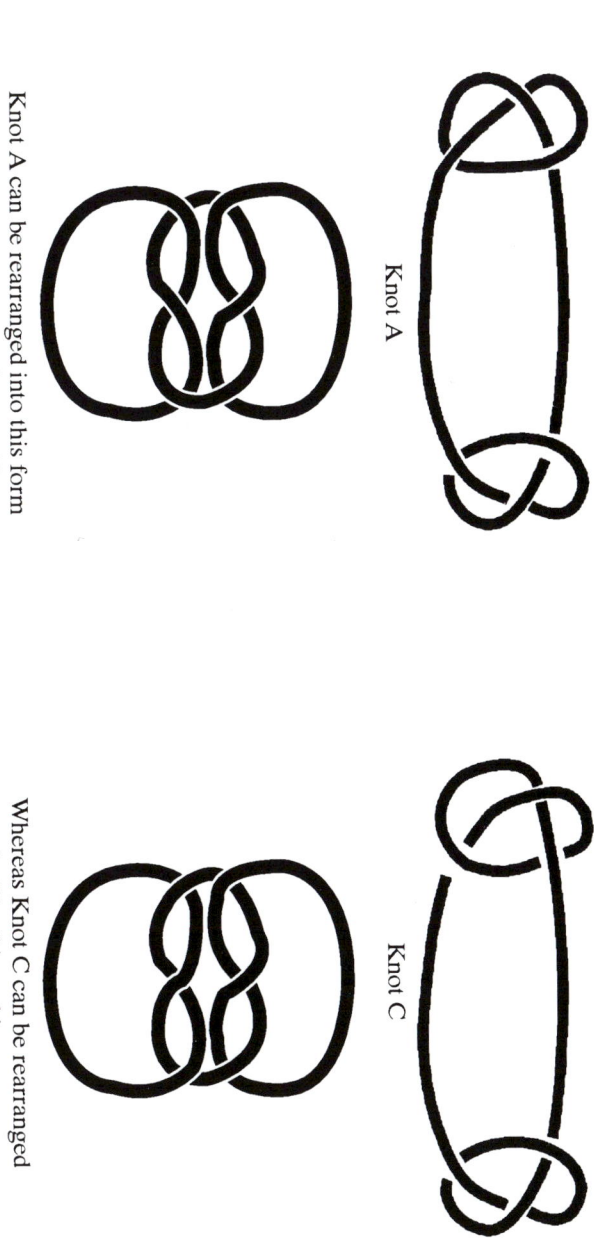

Knot A

Knot C

Knot A can be rearranged into this form and can be clearly seen to be a reef knot.

Whereas Knot C can be rearranged to be like this and is a granny.

Both of these composite knots are made from two trefoils. The difference is whether they are identical or mirror images. If one trefoil is linked to its mirror image then the result is a reef knot. If two identical trefoils are linked, then the result is a granny.

Puzzle 10

Is it true that there are two distinct kinds of granny, but only one kind of reef knot?

Prime Knots

Knots like the trefoil, the figure-eight and knots B and D from page 14 are all examples of prime knots. It is not possible to split any of them up into smaller parts.

A prime knot is a knot that is not made up of simpler knots!

All the knots with crossing numbers of 3, 4 and 5 are prime. However, once we consider knots with crossing numbers of 6 and higher, they can be either prime or composite.

Puzzle 11

Which of these knots are prime and which are composite?

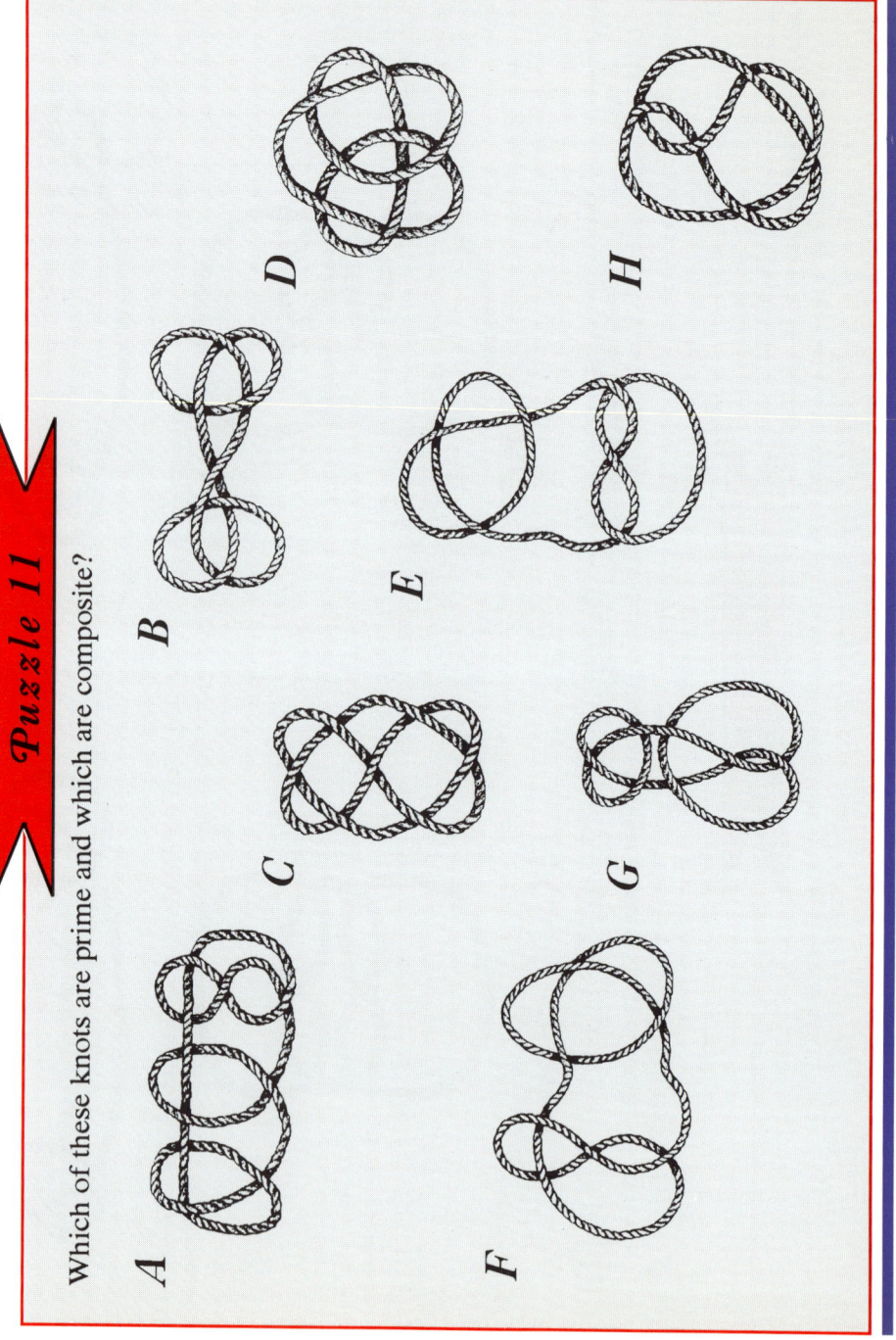

Adding Knots

Since composite knots are made from simpler knots we can now think about what happens when we add two or more knots together. The addition of knots behaves rather like the addition or multiplication of numbers, where it does not matter which number comes first.

As an analogy with the facts that 6 + 3 = 3 + 6, and 6 x 3 = 3 x 6, it does not matter which knot is tied first. The resulting composite knot is the same.

So a trefoil plus a figure-eight knot is exactly the same as a figure-eight knot plus a trefoil. In this case it is also true that the crossing numbers also obey the same commutative rule.

$$4 + 3 = 7 = 3 + 4$$

So far, knot theorists have been unable to prove conclusively that this rule for the addition of crossing numbers is true for all composite knots. However, no examples have yet been found where it is not true.

The use of the word 'prime' in prime knots suggests a further analogy with prime numbers and multiplication. Just as a prime number is only divisible by itself and 1, so a prime knot is a knot that can only be composed of itself, or itself plus any number of unknots.

The 3-Colour Test

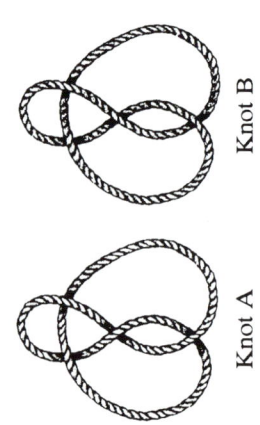

Knot A   Knot B

"Are these knots the same?" is a fundamental question in knot theory. "They look the same" could be the response when looking at the pair of knots on the left.

However, by rearranging them physically or mentally you would soon discover that they do not have the same crossing number. The crossing number of A is 4 and of B is 3 and we are therefore certain that they are different.

Rearranging and then using crossing numbers to distinguish between knots may be possible when dealing with simple diagrams but we need other methods which we can use for more complicated knots. One useful method is provided by what is called 'the 3-colour test'. In a similar way to the crossing number, '3-colourability' is an invariant property of a knot and remains unchanged no matter how much the knot is rearranged. It is a test which enables us to prove that two knots are different. But, like the crossing number, it cannot tell us that they are definitely the same.

The technique is to colour the knots using exactly three colours while obeying the following rules.

THE 3-COLOUR RULES

1. **Every strand must remain the same colour between two neighbouring under-crossovers.**
2. **There must be either one or three colours meeting at each crossover.**
3. **Crossovers with two colours meeting are not allowed.**

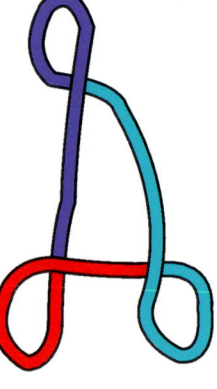

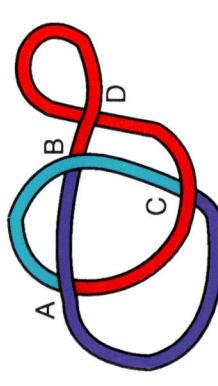

There are three colours at crossovers A, B, & C, and one at crossover D. Three colours have been used and all the rules have been obeyed.

Although the diagram is coloured with three colours there are only two colours at each crossover and so the rules have been broken.

Let us now attempt to 3-colour the knots A & B opposite.

Knot A is not 3-colourable.
To maintain the pattern of three colours at a crossover either two strands have to change colour between two under crossovers or the diagram has to be coloured with four colours.
Neither is permitted under the rules of 3-colourability.

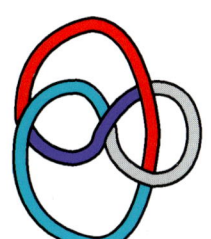

Knot B is 3-colourable.
There are three colours at each crossover and exactly three colours have been used.

This result confirms what we already know, that knot A and knot B are different.
To test in general that two knots are definitely the same is a more difficult problem and requires more advanced knot theory. See the reading list on page 42.

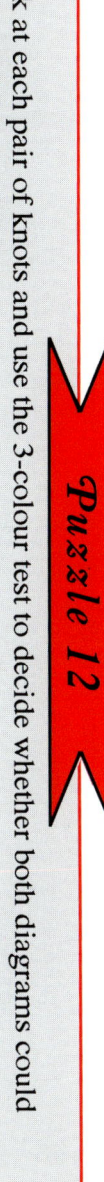

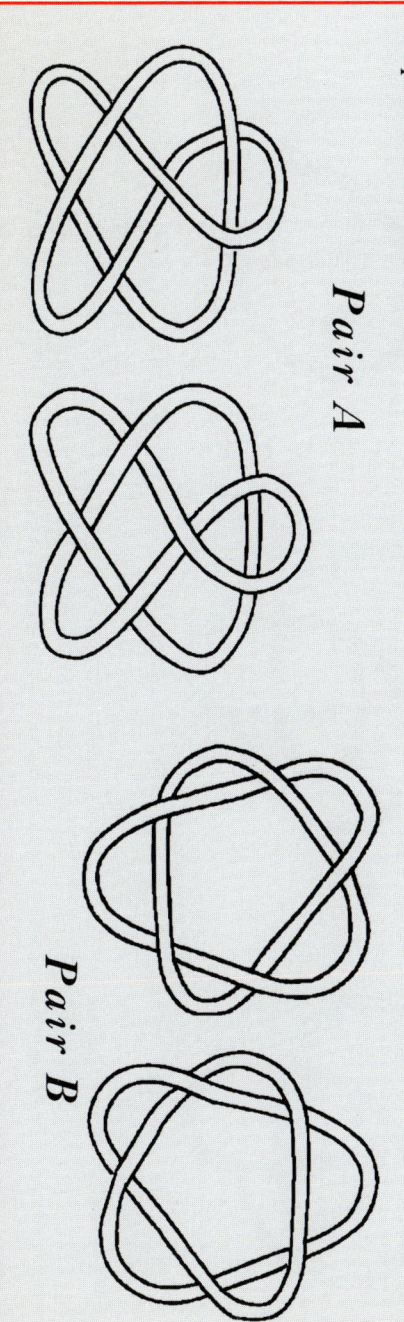

Pair A

Pair B

Puzzle 12

Look at each pair of knots and use the 3-colour test to decide whether both diagrams could represent the same knot.

A Strategy for 3-Colouring

When applying the 3-colour test it is necessary to have a systematic approach. If at first you do not succeed, it does not mean that the knot is not 3-colourable. Try again! You must work in an orderly and methodical way so as to consider every possibility at every crossover.

Let us test to see if this knot passes the 3-colour test.

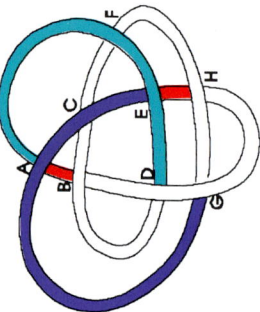

We start by first labelling all the crossovers.

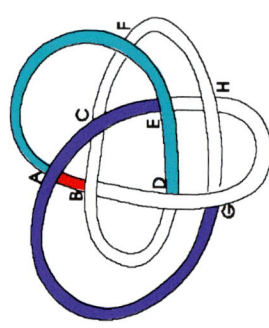

Then we colour the three sections at A. Each is coloured as far as the next under-crossover.

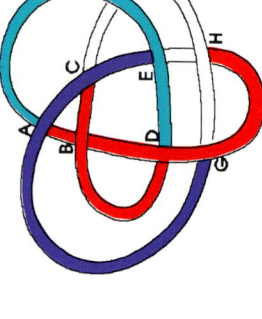

Then we look for a crossover that is nearly complete and colour it. Here E - H must be red.

There are no further crossovers which are nearly complete, so we must choose one which is partially coloured and test to see what happens when it is coloured with either one or three colours. We shall attempt to colour the strands at crossover B.

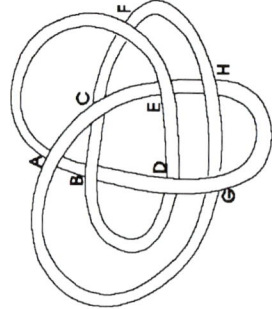

If C - D is green
then B - H must be purple.
This would mean that there are
only two colours at D. Hence this
approach has to be abandoned.

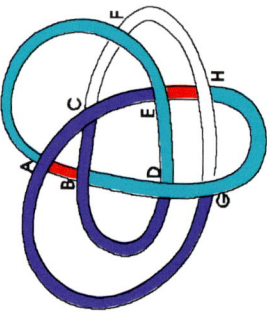

If C - D is purple
then B - H must be green.
This would then require that
there are only two colours at D
and this strategy also fails.

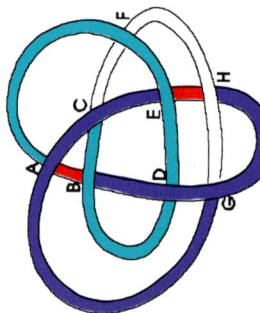

There is another possibility.
Perhaps both C - D and B - H are red.
However, then there would have to be
only two colours at D and
this is not permitted.

At this stage it would be easy to conclude that this knot fails the 3-colour test. However, there is still one further option to try. We had assumed that there were three colours at A. Suppose that this assumption was wrong and that there is only one colour at that crossover?

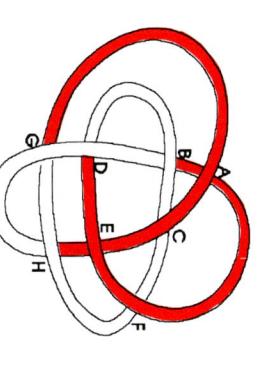

Let us colour all three sections at A red. We must then colour E - H red to complete crossover E.

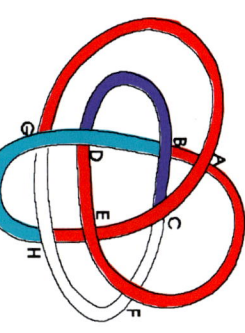

There are no nearly complete crossovers to tackle next, so we try to colour the sections at B.

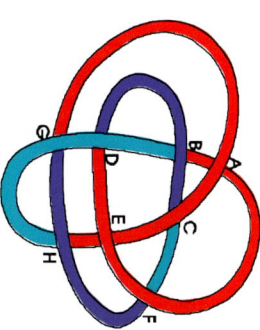

It is now possible to successfully complete the diagram. This knot does pass the 3-colour test.

It is only when every logical combination of colouring has been tried and all have proved impossible that it can be said that a knot has failed the 3-colour test.

Use a similar strategy to discover which of these knots pass the 3-colour test.

A   B   C

Puzzle 13

Analysing Moves

Manipulating a knot in order to simplify it or put it into a more easily recognisable form means moving strands and creating or destroying crossovers. We do this instinctively, but in 1929 the knot theorist Reidemeister set out to analyse exactly how many different moves were possible. He concluded that there were only three! Four, if you take a certain point of view.

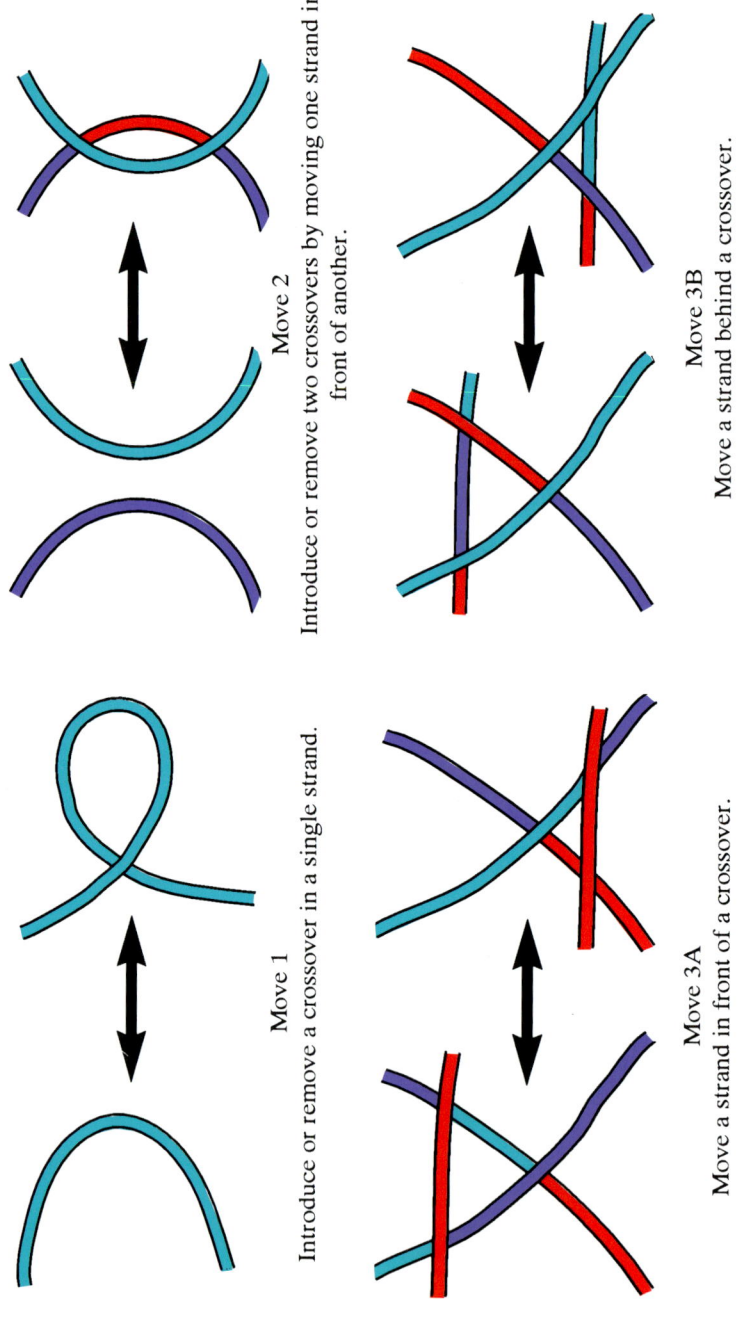

Move 1
Introduce or remove a crossover in a single strand.

Move 2
Introduce or remove two crossovers by moving one strand in front of another.

Move 3A
Move a strand in front of a crossover.

Move 3B
Move a strand behind a crossover.

Strictly speaking, moves 3A & 3B are identical, because a knot can be looked at from either side. All four diagrams have been 3-coloured in order to demonstrate that none of the moves disturbs whether a knot passes or fails the 3-colour test. Note how the colours of the cut ends of each strand are unchanged by the move. Since any process of rearrangement is simply a sequence or combination of these moves, these diagrams serve as a proof of the invariance of 3-colourability.

Puzzle 14

Each of these pairs of knots has been rearranged using one of the Reidemeister moves opposite. Which move was used for each pair?

Pair A

Pair B

Pair C

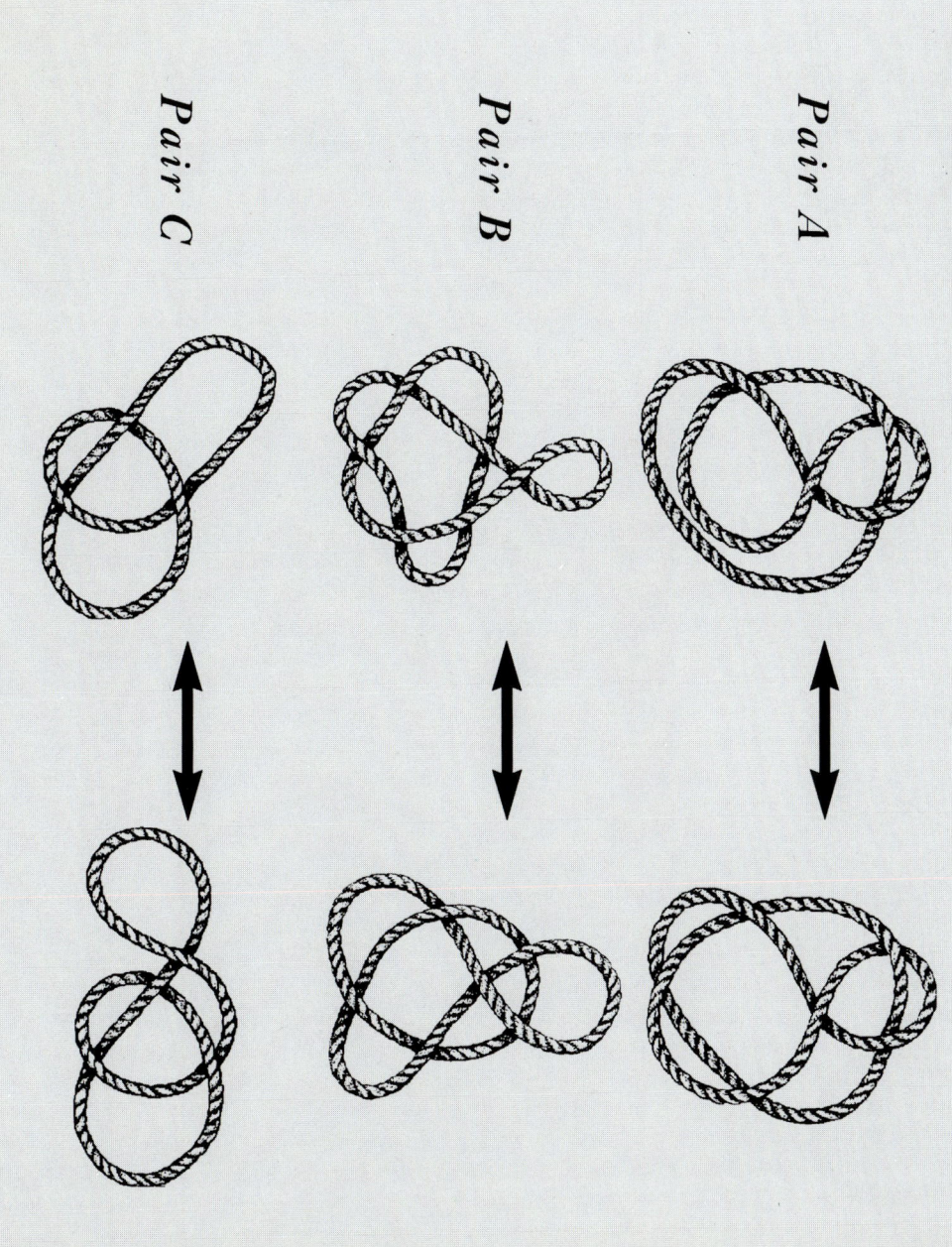

3-Colouring Composite Knots

Here are four prime knots, two of which pass the 3-colour test.

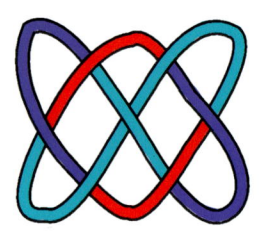

Knot A: The trefoil
(passes the 3-colour test)

Knot B: The figure-eight
(fails the 3-colour test)

Knot C: The pentoil
(fails the 3-colour test)

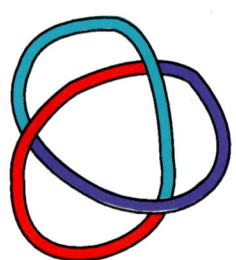

Knot D
(passes the 3-colour test)

Puzzle 15

These composite knots have been made from the knots above. Which of them passes the 3-colour test?

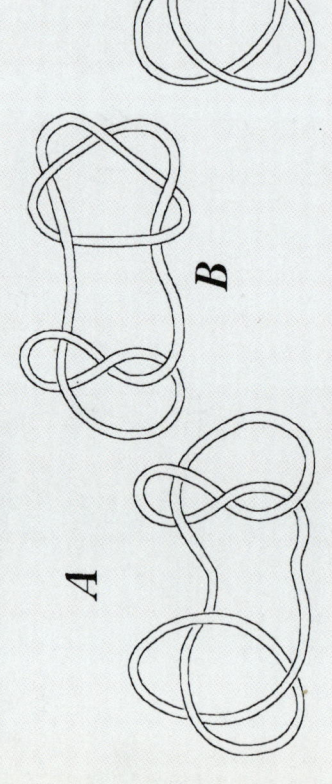

A   B   C   D

Puzzle 16

Using the result of the last puzzle, is there a rule which could be applied to produce a whole range of composite knots that always pass the 3-colour test?

Crossing Number 6

There are exactly three distinct prime knots which have a crossing number of 6.

Only the middle one passes the 3-colour test.

Puzzle 17

Which of these knots are 3-colourable and are therefore rearrangements of the middle knot above?

A

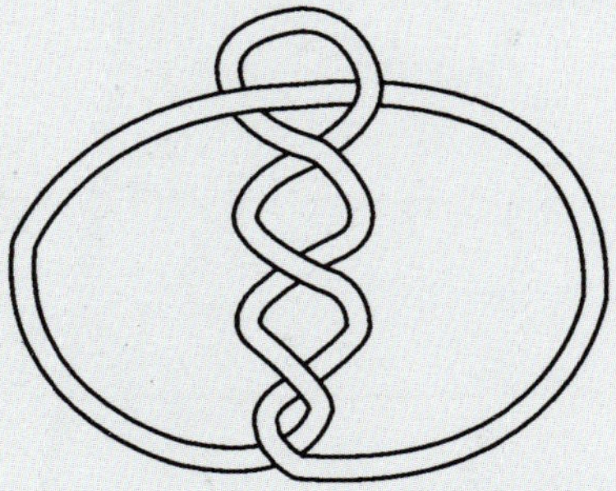

B

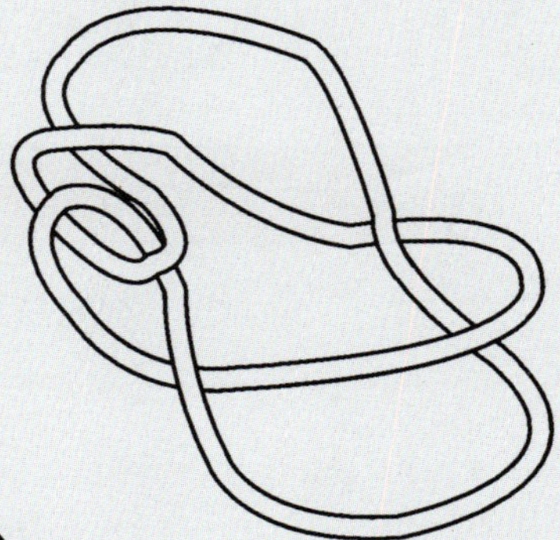

C

D

Crossing Numbers 7+

There are exactly seven distinct prime knots which have a crossing number of 7.

Puzzle 18

There are also two composite knots with a crossing number of 7. What are they?

We have now learned a lot about knots and have had to solve some intriguing puzzles on the way. However as this table shows, there are still plenty of different knots to study!

PRIME KNOTS

Crossing number:	0	1	2	3	4	5	6	7	8	9	10	11	12	13	...
Number of prime knots:	1	0	0	1	1	2	3	7	21	49	165	552	2176	9988	...

Some more advanced mathematical ideas are discussed in the section beginning on page 35, but for the moment let us look for some further puzzling notions amongst the sort of knots that people use in the practical and artistic world.

Section 2

The Knots Puzzle Book

Practical & Artistic Knots

Contents ...

Bends	28
Hitches	30
Decorative Knotwork	32
Stitched Designs	34

Bends

A 'bend' is a knot that is employed to join two ropes together. The reef knot is probably the best known example and it is shown here with four other bends which are in common use.

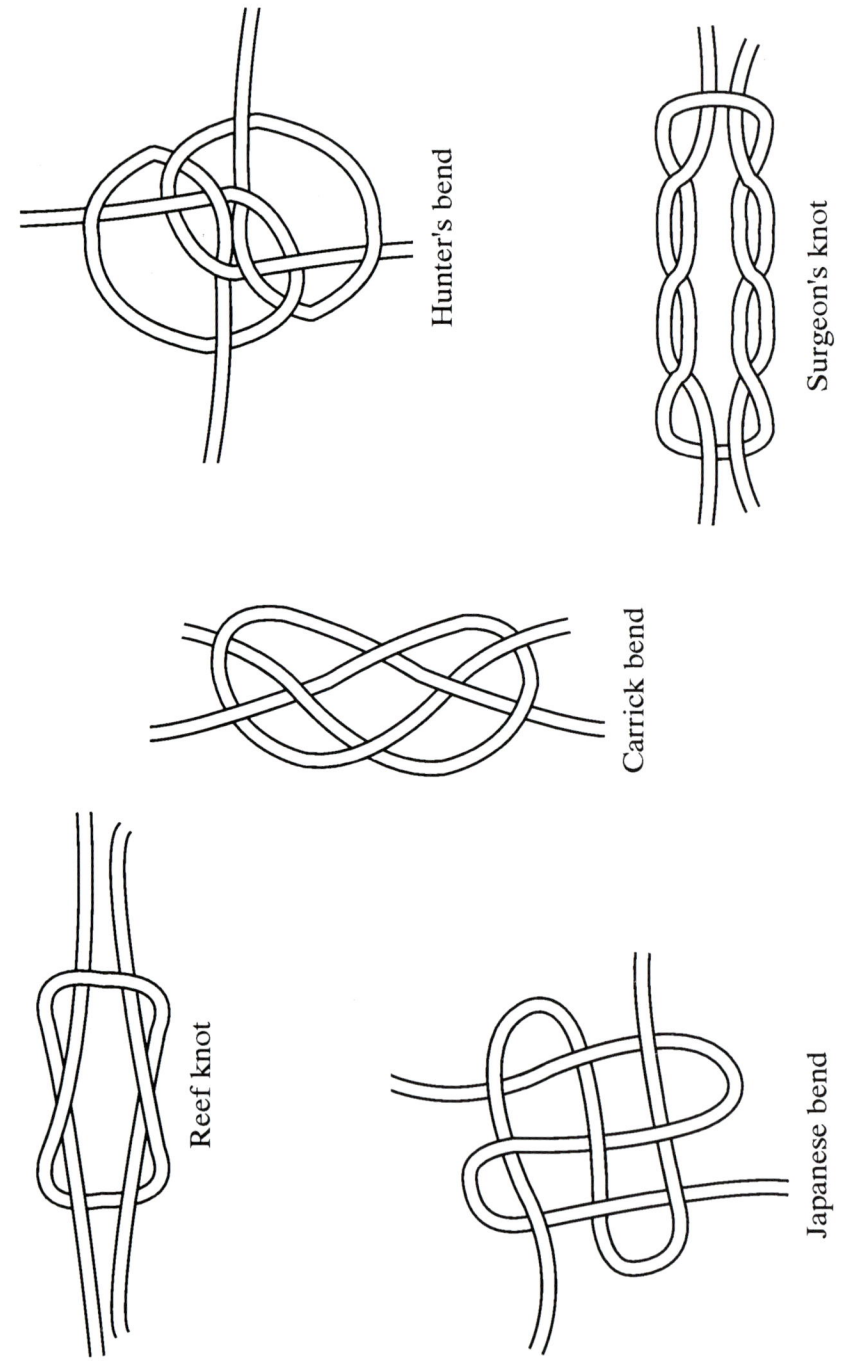

Reef knot

Hunter's bend

Carrick bend

Surgeon's knot

Japanese bend

Puzzle 19

In the Carrick bend, each rope crosses itself as well as crossing the other. In which of the bends illustrated above does each rope never cross itself?

Puzzle 20

Which of these knots has essentially the same structure as the reef knot and the surgeon's knot?

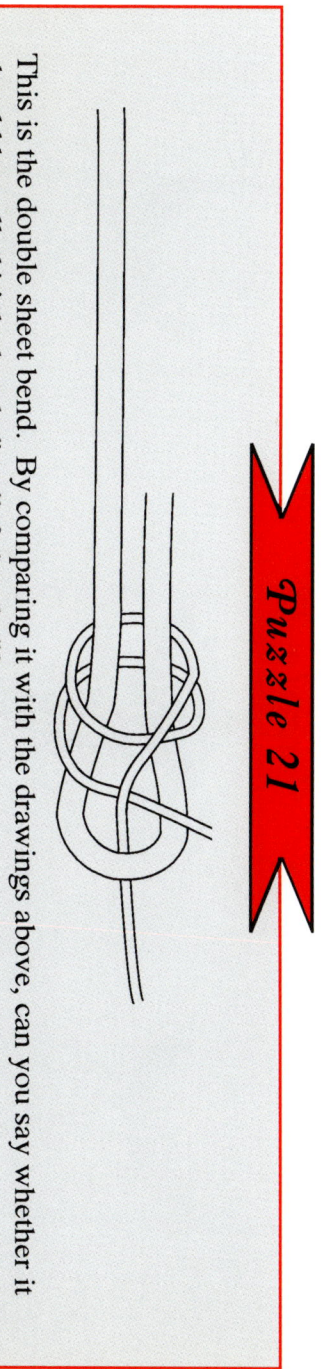

A   B

Bends are most secure when they are used to join ropes of equal thickness. However, the sheet bend is an exception. It is one of the few knots which is effective in joining ropes of differing diameters.

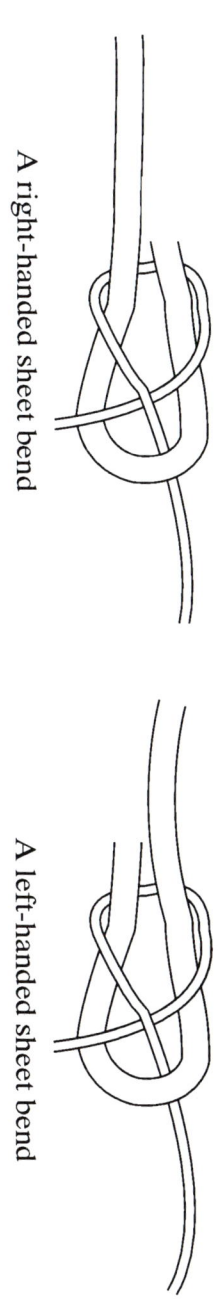

A right-handed sheet bend   A left-handed sheet bend

Note the position of the shorter end of the thicker rope for each of these versions of the sheet bend. The left handed version is a much less secure knot than its right handed counterpart.

Puzzle 21

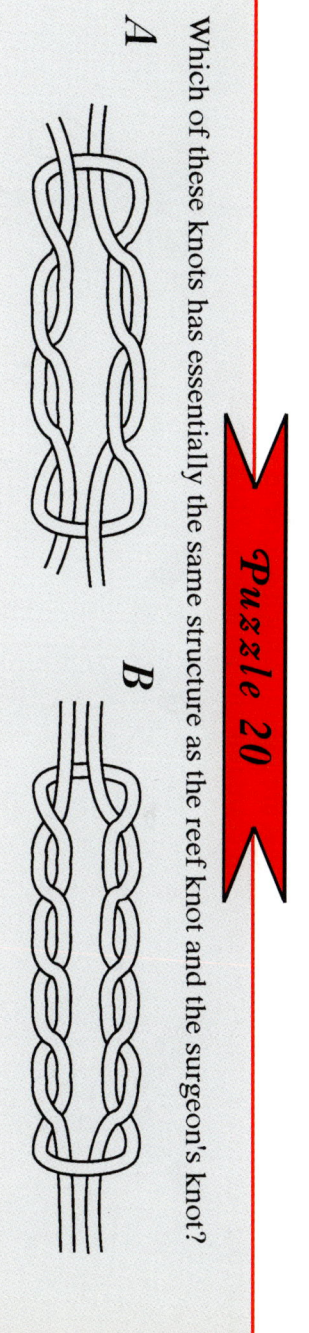

This is the double sheet bend. By comparing it with the drawings above, can you say whether it should be called 'right-handed' or 'left-handed'?

Hitches

A 'hitch' is the name given to knots that are used to attach a rope to something solid such as a rail, a post or a ring. Here are three examples of common hitches attached to a ring.

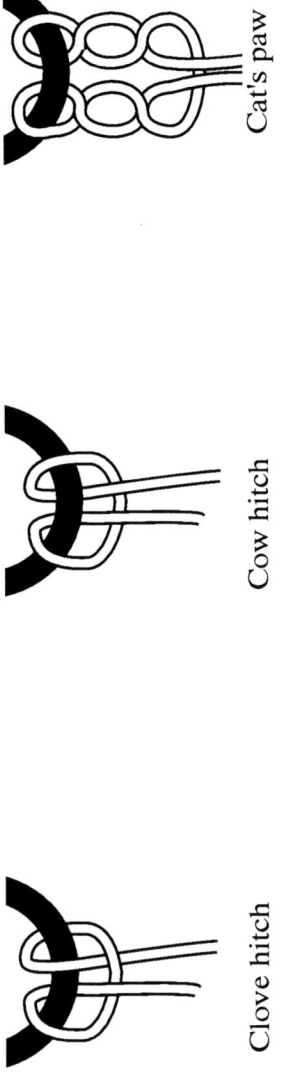

Clove hitch   Cow hitch   Cat's paw

Careful examination will show similarities and differences. They are all similar in that two strands pass through the ring. The clove hitch and the cow hitch each have two crossovers below the ring, whereas the cat's paw has six. The only difference between the clove hitch and the cow hitch lies in the treatment of the loose ends. They are either both together through the loop or one in and one out.

Puzzle 22

Is the cat's paw more like the clove hitch or the cow hitch?

Puzzle 23

Explain which of these knots is the odd one out.

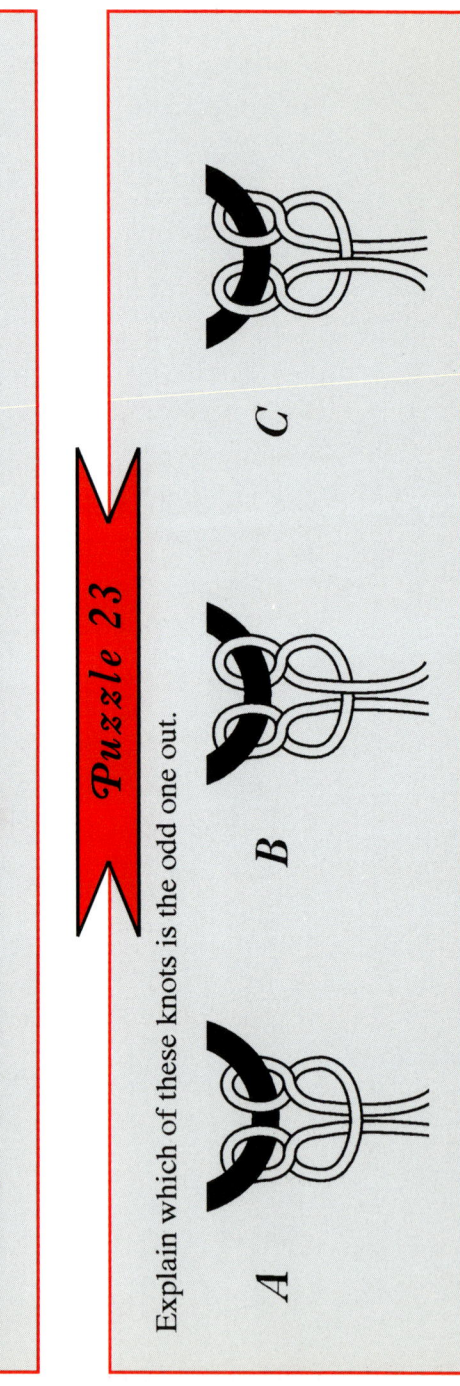

A   B   C

Remembering how to tie a particular knot is often made easier if you first learn a set of verbal instructions. For instance, the classic way of remembering how tie a reef knot is to say to yourself "Right over left and under and then left over right and under". Below are the instructions for tying half hitches.

One half hitch
"Under, over and through"

Two half hitches
"Under, over and through and then under, over and through"

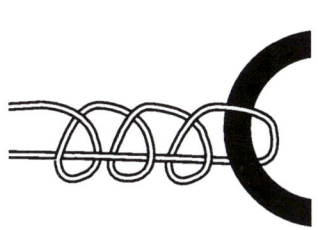

Three half hitches
"Two half hitches and then over, under and through"

Puzzle 24

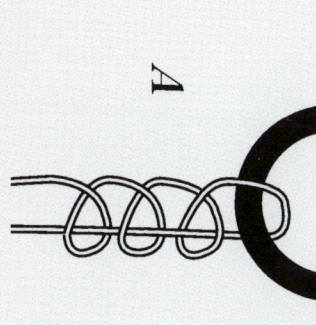

A

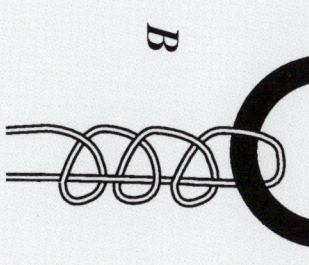

B

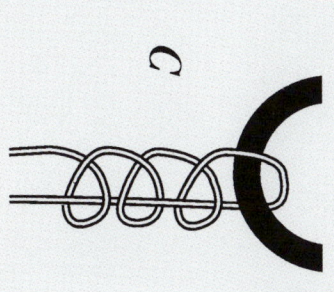

C

Which of these knots was tied following the pattern, "under, over, through, then over, under, through, and finally under, over through"?

Decorative Knotwork

Knots have long been a powerful source of artistic inspiration as well as being valued for their more obvious practical uses. Such designs are often referred to as 'Celtic patterns' because the Celts often decorated their manuscripts, jewellery and stonework with complicated knot patterns. However this style of decoration has also been used by many other cultures throughout history.

Puzzle 25

Which 'mathematical' knots are these patterns equivalent to?

Puzzle 26

These two knot patterns may look similar but what is the fundamental difference between them?

Puzzle 27

How many separate strands are needed to make up each of these designs?

A

B

C

Stitched Designs

Stitching, knitting, weaving, and sewing are all processes where one or more lengths of yarn are knotted or linked together to produce useful articles. They also suggest further attractive designs for drawing.

Puzzle 28

Is it true that one or both of these designs is really an unknot?

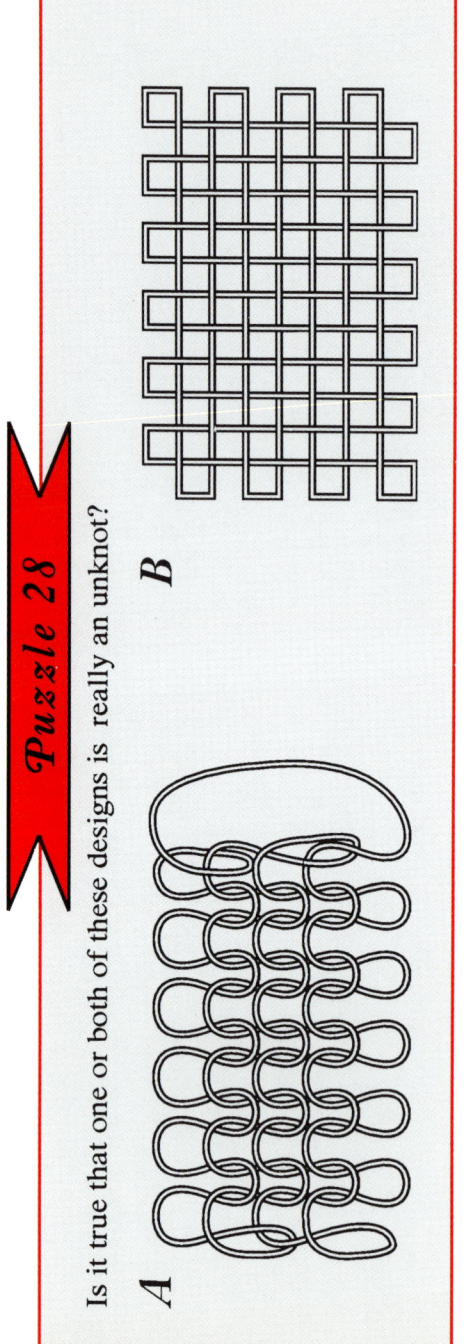

A

B

Puzzle 29

Chain stitch is often used to sew up the top of potato bags. Which end of the string should you pull to open the bag?

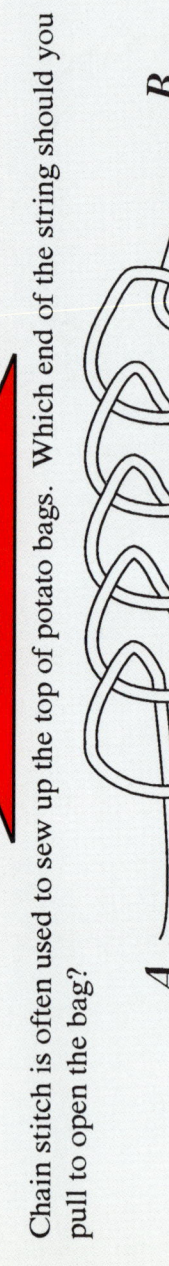

A

B

Of course there are many other beautiful knotted structures and patterns to be discovered in everyday life. If you keep your eyes open, you will find them in lots of different places.

The next section introduces some more advanced mathematical ideas to help with classifying and understanding how knots can form families and how they can link together.

Section 3

The Knots Puzzle Book

More Advanced Mathematical Ideas

Contents ...

Knot Families	36
Links	38
Are there More Colour Tests?	40

Knot Families

On pages 28-31, certain knots were grouped together as 'Bends' or 'Hitches'. They were classified into 'families' of related knots, in those instances by their practical function. Knots can also be grouped into families according to their geometry or general shape. For instance, the figure-eight family below.

Puzzle 30

Here are six members of the figure-eight family, although not all are shown arranged into standard figure-eight form. Try making D, E, and F and then rearranging them. The first member has a crossing number of 4. Do all members of the family have an even crossing number?

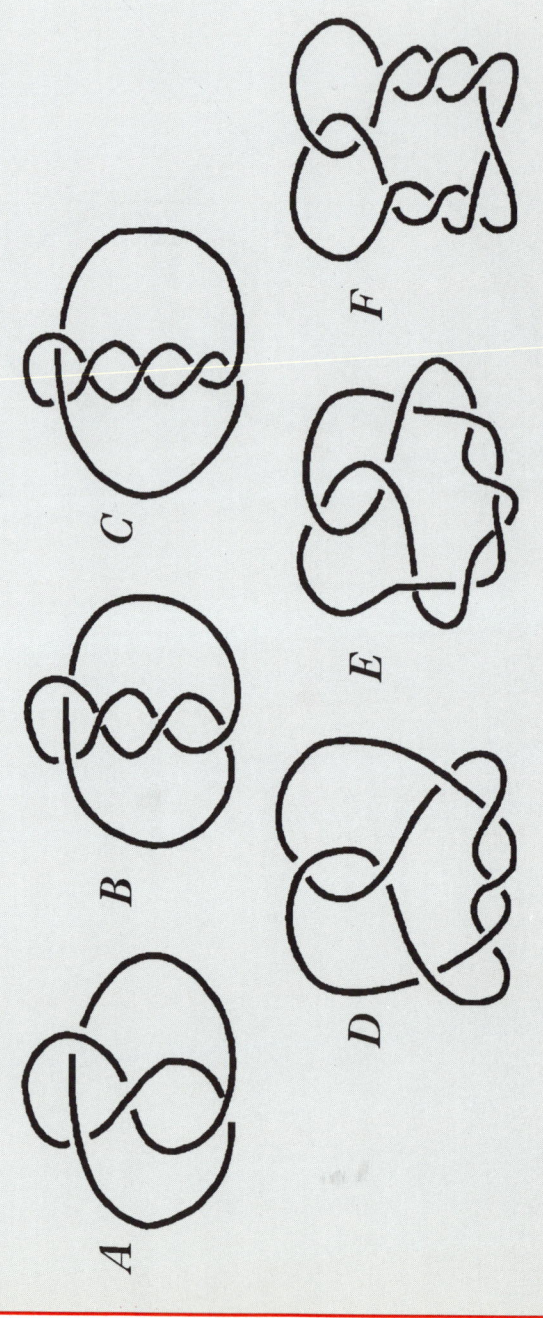

Puzzle 31

Is it true to say that none of the knots in this family can be 3-coloured? Are all the knots in this family prime?

Another interesting family of knots is called the 'torus' family. These are knots which will fit neatly and snugly around the outside of a doughnut shape or torus.

Within the torus family there is a lesser grouping which is known as the 'overhand knot' family. Its simplest member is the trefoil shown on the left. All of its members can be generated from the middle diagram by simply adding more and more twists to replace the dots across the centre. Two extra twists gives the pentoil, four the septoil, six the nonoil and so on. Members of the torus family can have even or odd crossing numbers, but all the members of the 'overhand knot' section of it have odd crossing numbers.

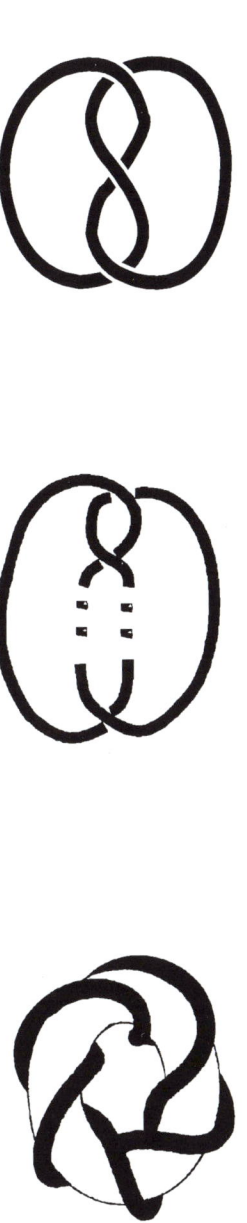

Puzzle 32

Here are the first four knots in the 'overhand knot' family. Which pass the 3-colour test?

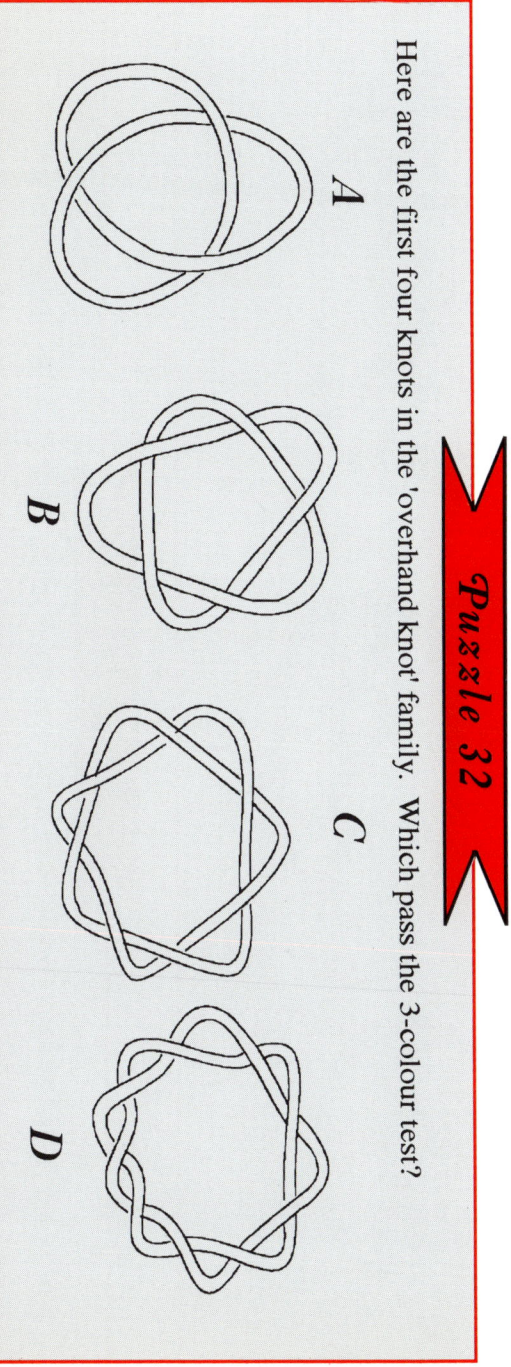

A   B   C   D

Links

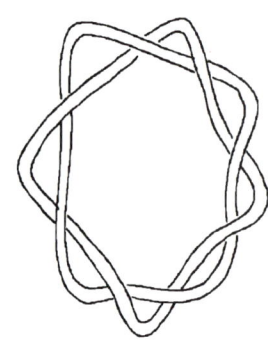

At first sight, this diagram might seem to represent a knot in the overhand knot family described overleaf. However a count will show that there are six crossovers, an even number and not the expected odd number. It is a torus knot but careful examination will reveal that it is not a simple knot made of a single strand, but one constructed from two distinct linked strands.

It is clear that an attempt to classify various patterns of linked ropes provides another topic worthy of investigation. These simple puzzles could suggest various starting points for such an investigation. Do remember that it is usually best to study knots by actually tying them as well as by drawing them.

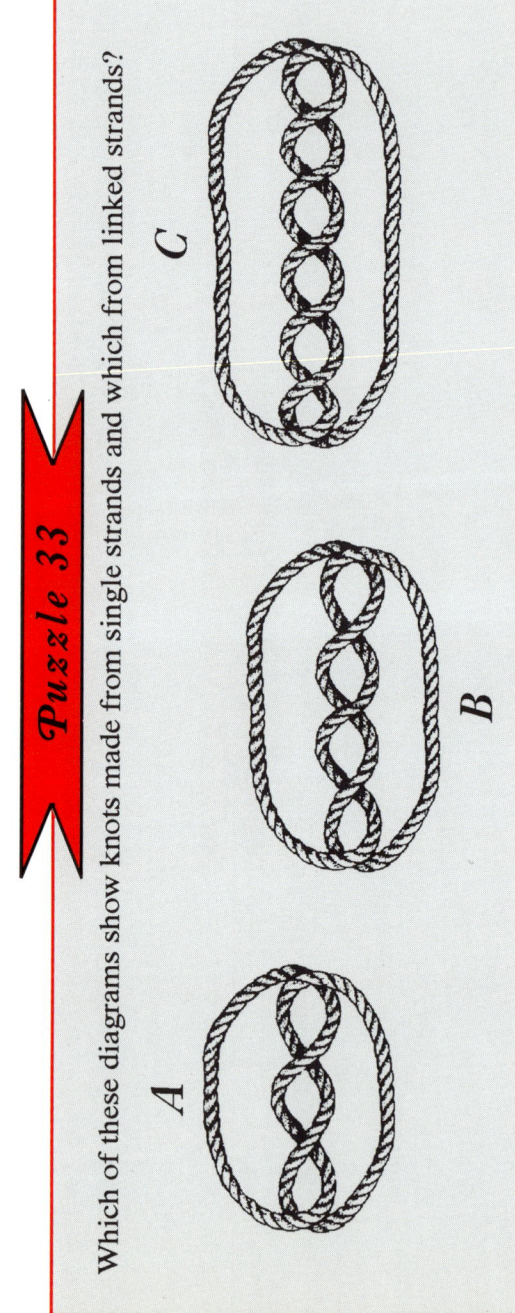

Puzzle 33

Which of these diagrams show knots made from single strands and which from linked strands?

Puzzle 34

How many linked strands do you need to make a ten crossing version of the diagrams above?

A good way of investigating more complicated links is to make use of colour, either by colouring in a knot diagram or tying the links with differently coloured ropes.

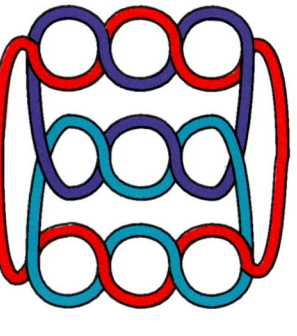

Puzzle 35

Here are some examples of more complicated links. How many separate strands are there in each?

A

B

C

Puzzle 36

Is there an easy way to predict how many strands knots like these will have?

Are there More Colour Tests?

The 3-colour test is a technique which is able to distinguish between certain pairs of knots but one with a rather limited usefulness. All it can really do is to divide knots up into two classes, those which pass the test and those which fail. This is rather like deciding whether numbers are even or odd. It tells us something about the knot, but not much. We therefore wonder if there are other similar tests which are better able to distinguish between different knots. For instance is there a 4-colour test, a 5-colour test or perhaps a whole series of such tests?

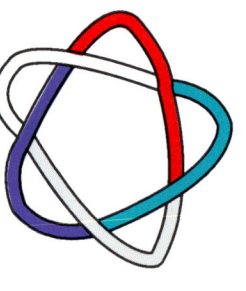

In a similar way a pentoil knot can be coloured with five colours.

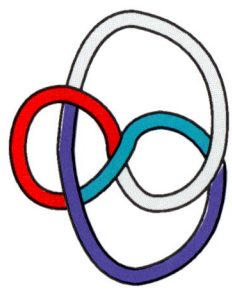

Retaining the rule that one or three colours must meet at each crossover, a figure-eight knot can be coloured with four colours.

With just two examples it would be easy to assume that both the 4-colour and 5-colour tests do exist and that each obeys the same rules as the 3-colour test. Knot enthusiasts have studied this problem in detail and have discovered that the 4-colour test does not work, but that the 5-colour test does. In fact it is just the start of a series of n-colour tests where n is always an odd number. No-one has yet found a knot which satisfies the 9-colour test and it is thought that n has to be a prime number.

Once more than 3 colours are used, it becomes difficult to keep track of them and it is usual to label the sections with numbers instead, using 0, 1, 2, 3, 4 etc.

The diagram on the right shows the trefoil knot '3-coloured' by being labelled with the three 'mod 3' numbers 0, 1, 2.

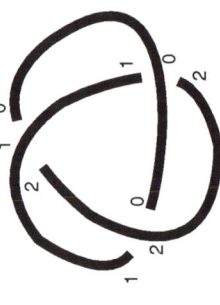

From this idea grows a whole new study of knots using 'modulo' arithmetic and new ways of classifying knots into groups and types. On page 46 there is a list of books for those who would like to carry an investigation of this approach to knot theory further.

Section 4

Knot Classification

There are only 36 distinct prime knots which have crossing numbers of 8 or less and they are all listed on the next four pages. Any prime knot you can tie with a crossing number in that range will be there, together with its unique classification code. The list does not include mirror images. As has already been shown some knots do have distinct mirror images whereas others do not. Take this into account when trying to classify a knot that you have tied. Consider also the possibility that your knot is not prime, but composite.

These classification pages may be photocopied for personal or classroom use, but not for commercial purposes.

Knots Classification

Prime Knots with Crossing Numbers 0 - 6

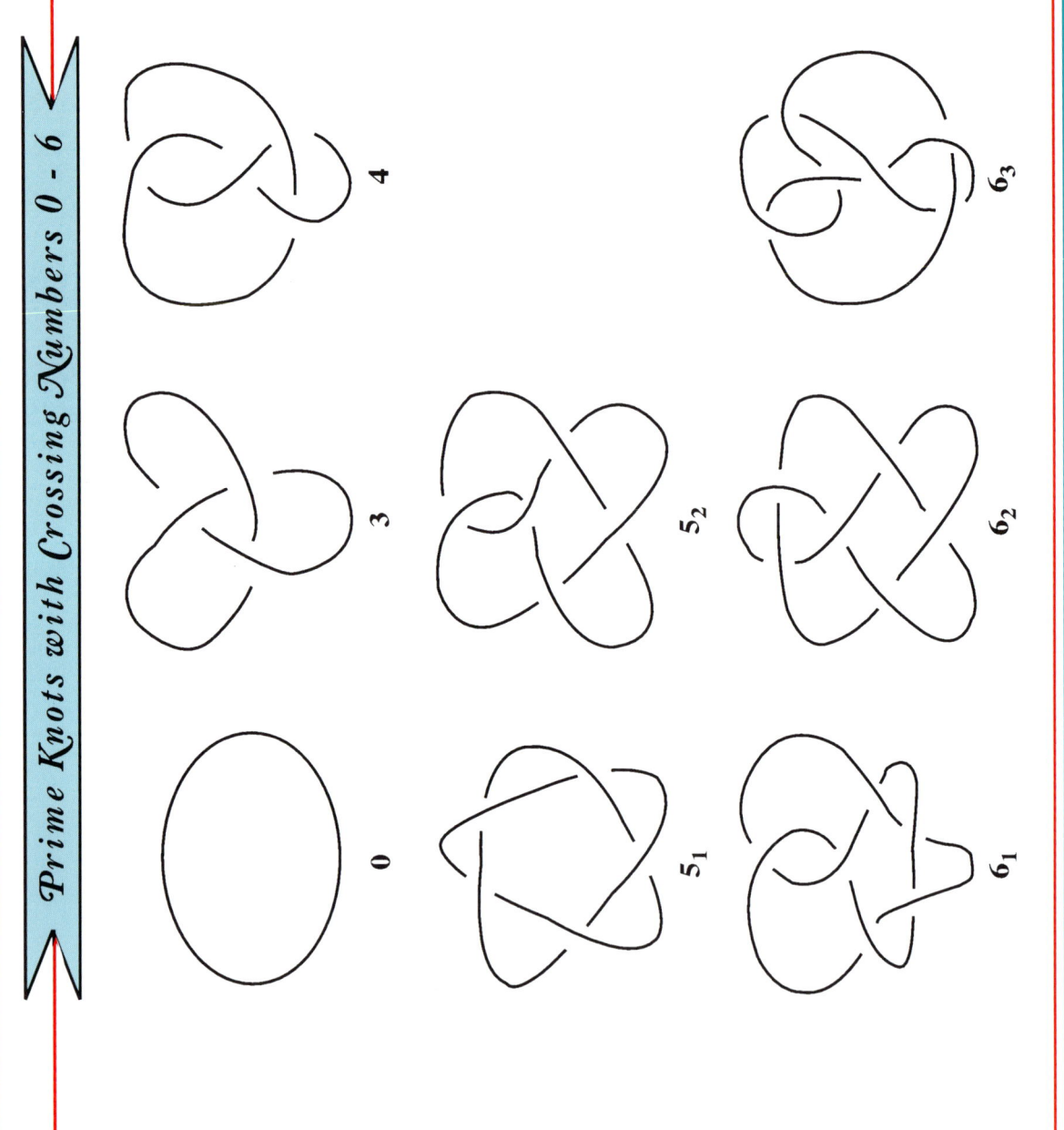

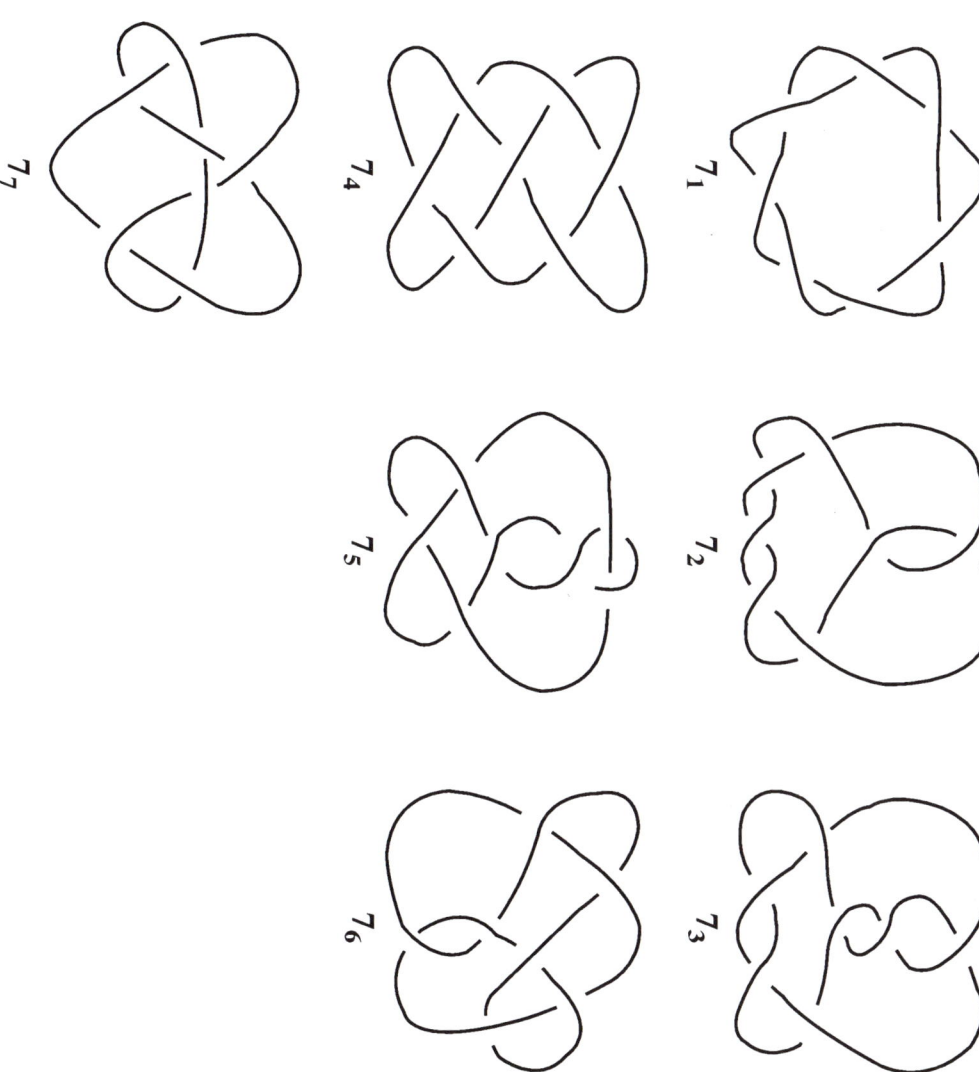

Knots Classification

Prime Knots with Crossing Number 8

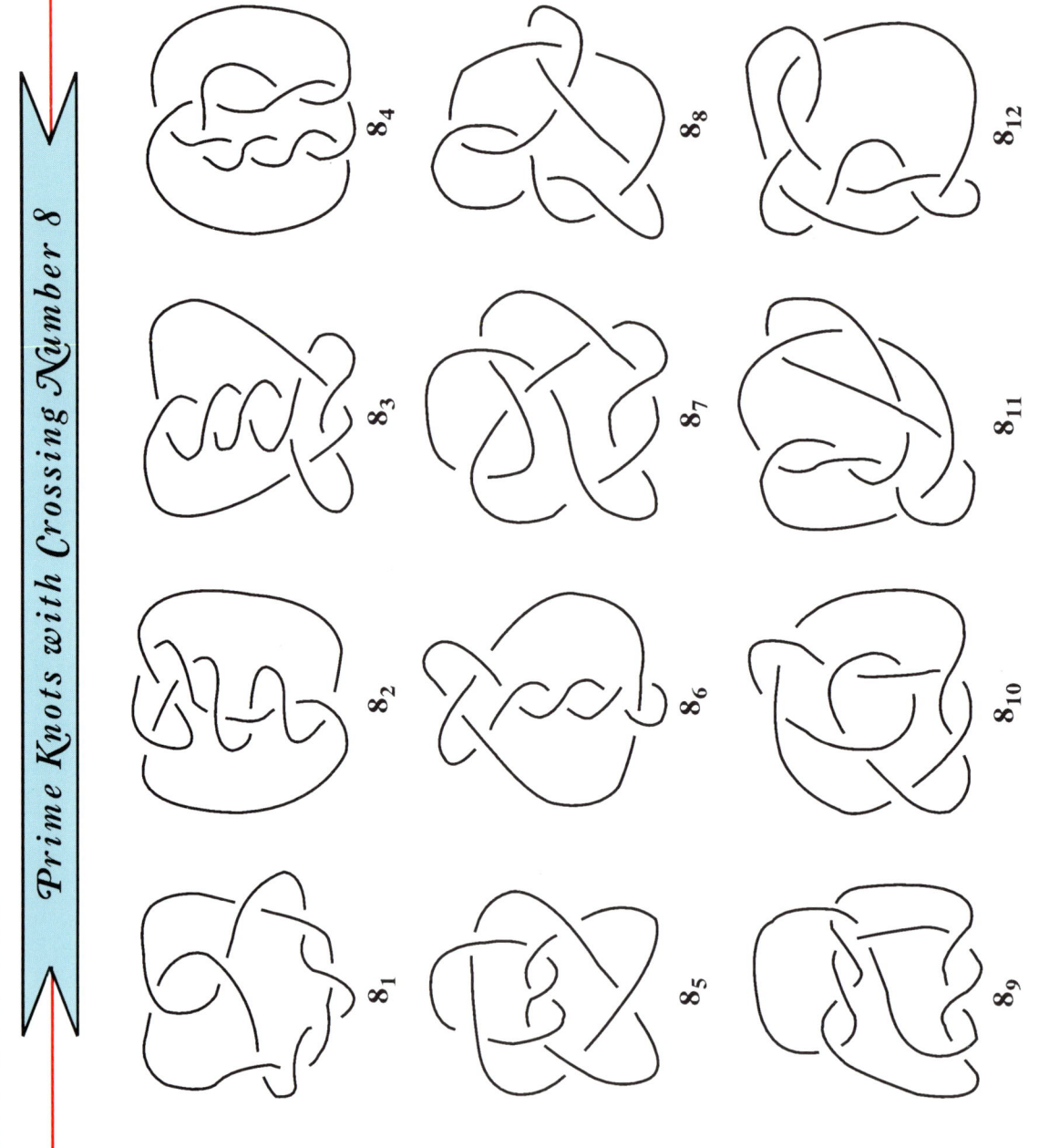

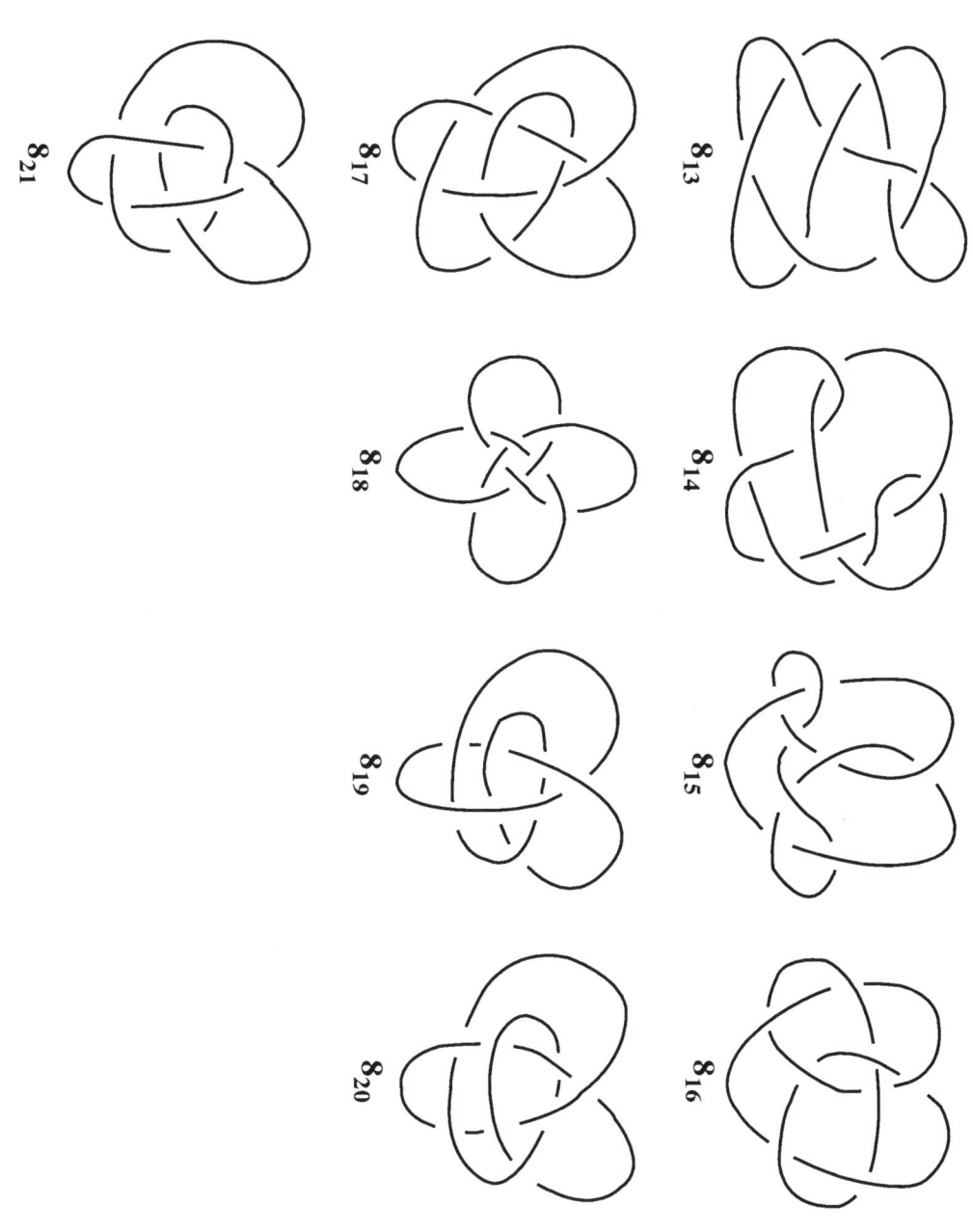

Ever Onwards ...

I should particularly like to thank Professor Ronnie Brown who first aroused my interest in knots and Dr Peter Cromwell, Dr Tim Porter and Peter Hines who kindly answered my many questions and enquiries.

It is hoped that the puzzles and ideas in this book have been interesting and stimulating to solve. Perhaps they suggest other knot ideas to think about and investigate?

Below there are four books which will be of interest to anyone who would like to find out more about knot polynomials, n-colouring using modulo arithmetic and other more advanced aspects of knot theory.

The study of knots is as endless as the mathematical knot itself!

More about Mathematical Knots

Gilbert, N.D. and Porter, T. *Knots and Surfaces* (Oxford University Press 1994)

Kauffman, L.H. *On Knots* (Princetown University Press 1987)

Kauffman, L.H. *Knots and Physics* (World Scientific 1991)

Rolfsen, D. *Knots and Links* (Publish or Perish Inc. 1976)

Section 5

The
Knots
Puzzle Book

Solutions
&
Comments ...

Solutions to the Puzzles

1. A and D are actual knots.
2. A has no real crossings.
3. B, C, D, F, G and H are unknots.
4. A has crossing number 6, B has crossing number 0, C has crossing number 4.
5. The crossing numbers are
   A: 5   B: 4   C: 3   D: 6   E: 0   F: 8.
6. Only B is a figure-eight knot.
7. C, E and H are unknots, A, D, G and I are trefoils, B and F are figure-eights.
8. A, B and C are pentoils.
9. A and C.
10. Yes.
11. C, F and H are prime.
    A: 3+3+4   B: 3+3   C: prime 7   D: 3+3
    E: 3+3   F: prime 3   G: 3+4   H: prime 5.
12. Pair A are both 3-colourable so could be the same knot. In fact they are both trefoils. Pair B are not the same, because the first is not 3-colourable while the second is.
13. All three are 3-colourable.
14. Pair A: move 2   Pair B: move 3B   Pair C: move 1.
15. A, C and D are 3-colourable.
16. Any composite knot which contains at least one 3-colourable knot is itself 3-colourable by using a single colour for the remainder of the composite knot.
17. A and B.
18. Anticlockwise trefoil + figure-eight and Clockwise trefoil + figure-eight.
19. Reef knot, Japanese bend and Surgeon's knot.
20. B.
21. Left handed.
22. Cow hitch.
23. A.
24. C.
25. A: figure-eight   B: trefoil   C: link with four crossings and the unknot.
    This is sometimes called the 'King Solomon's' knot.
26. A has 5 strands, B has only one.
    If a knot pattern of this kind is based upon a grid whose dimensions are co-prime, then there is only one strand.
27. A and C are each made up of 3 strands.
    B is made up of 2 or 4 strands, depending on how the diagram is interpreted.
28. A is unknotted and B is knotted.
29. A.
30. No.
31. No, C and F are 3-colourable. They are all prime.
32. A and D.
33. B and C are simple knots, while A is a link.
34. 2.
35. A: 2   B: 1   C: 1.
36. It is the pattern of the number of crossings in each braid which determines the number of strands.
    2 braids: odd + odd = 2 strands   odd + even = 1   even + even = 2
    3 braids: odd + odd + odd = 1   odd + odd + even = 1   odd + even + even = 2   even + even + even = 3
    The order of odd and even does not affect the number of strands.